F9F Cougar

Written by Ken Neubeck

Walk Around®

Squadron Signal® Publications

Cover Art by Don Greer

About the Walk Around® Series

The Walk Around® series is about the details of specific military equipment using color and black-and-white archival and photographs of in-service, preserved, and restored equipment. *Walk Around®* titles are devoted to aircraft and military vehicles. These are picture books focus on operational equipment, not one-off or experimental subjects.

Squadron/Signal Walk Around® books feature the best surviving and restored historic aircraft and vehicles. Inevitably, the requirements of preservation, restoration, exhibit, and continued use may affect these examples in some details of paint and equipment. Authors strive to highlight any feature that departs from original specifications.

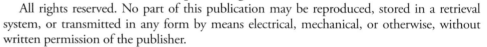

Proudly printed in the U.S.A.
Copyright 2012 Squadron/Signal Publications
1115 Crowley Drive, Carrollton, TX 75006-1312 U.S.A.
www.SquadronSignalPublications.com

Hardcover ISBN 978-0-89747-667-6
Softcover ISBN 978-0-89747-666-9

(Front Cover) A pair of early-production F9F-8 Cougars in Navy Blue paint scheme soar above the clouds.

(Back Cover) A photo-reconnaissance F9F-9P Cougar makes a pass over mountainous terrain.

Military/Combat Photographs and Snapshots

If you have any photos of aircraft, armor, soldiers, or ships of any nation, particularly wartime snapshots, please share them with us and help make Squadron/Signal's books all the more interesting and complete in the future. Any photograph sent to us will be copied and returned. Electronic images are preferred. The donor will be fully credited for any photos used. Please send them to the address above.

(Title Page) An F9F-8T Blue Angel aircraft is present at NAS Oceana in Virginia as part of the Blue Angel air show in 1964. The Blue Angels FTF-8T was one of the last F9F Cougar aircraft that flew actively in the United States. (Del Laughery collection)

Acknowledgments

The author wishes to thank Gary Chambers, Larry Feliu, John Gourley III, Clay Jansson, Bob Jellison, Del Loughery, Ray Neubeck, Ben Peck, Sal Picataggio, Adolfo Soto, Josh Stoff, Gary Verver, the Tailhook Association, the Northrop Grumman History Center and the Cradle of Aviation museum for their help in this project. This book was inspired by the photograph shown on page 30 that was taken by my dad when we attended a local air show in 1957. Cold War aircraft like the Cougar represent a part of U.S. history and it is important to preserve this legacy in photographs when opportunities exist.

Introduction

The F9F Cougar was the first swept-wing carrier-based fighter aircraft built for the U.S. Navy with the design based on the straight wing Grumman F9F Panther. In the original F9F Panther contract was a clause calling for design data for a swept-wing version. It was not until the Panther encountered swept-wing MiGs during the Korean War, however, that an urgent contract was signed in March of 1951 for a modified, swept-wing F9F Panther design.

Just six months after contract award, the first prototype of the F9F-6, the Cougar's first version, took to the air in September 1951. Although the aircraft was subsonic (except for dive maneuvers), the Cougar performed significantly better than the Panther. Initial production of the F9F-6 took place at the Grumman Bethpage, Long Island, facility from mid-1952 to mid-1954. The F9F-6 was equipped with a Pratt & Whitney J48-P-6 engine and four cannons in the nose, along with two pylons under the wing for either bombs or drop tanks. Initial production was for 646 F9F-6 aircraft, along with 60 additional F9F-6P reconnaissance version airplanes, with cameras fitted in the nose in lieu of the cannons.

The next production lot, the F9F-7, was equipped with the Allison J33 engine. However, the Allison engine proved to be less reliable and less powerful than the J48 engine and eventually most of the 168 F9F-7 aircraft that were built were converted to the J48 engine.

The next major version, the F9F-8 model, had an extended fuselage and increased wing area to accommodate larger fuel tanks and two additional pylons under the wing. The F9F-8 had four wing pylons to accommodate the AIM-9 sidewinder missiles. Production of 601 aircraft ran from early 1954 to early 1957. Some F9F-8 aircraft were converted to the F9F-8B, which had the capability of carrying nuclear bombs.

Another version of the Cougar was the F9F-8P photo reconnaissance model, of which 110 aircraft were produced. It had an extended nose section to accommodate 14 cameras.

There was also a two-seat trainer version of the Cougar, the F9F-8T, of which 399 aircraft were produced. A number of existing F9F-8 Cougars were converted to F9F-8B aircraft, which were used for low-altitude bombing.

The Blue Angels' fleet included 13 F9F-8 aircraft and two F9F-8T aircraft, beginning in 1954, with one F9F-8T Cougar remaining with the team until 1969.

The F9F-8 aircraft were withdrawn from front-line service by 1960 and were replaced by F11F and F8U aircraft. The only version of the Cougar to see combat were four F9F-8T Cougars that were used in the air command role, directing attacks against enemy positions during the Vietnam War in 1966 and 1967.

In 1962 the Navy launched a redesignation program in the course of which the F9F-8 became the F-9J, the F9F-8T became the TF-9J, and the F9F-8P became the RF-9J.

Two F9F-8Ts were inadvertently sold to the Argentine Navy in 1962, but saw only limited service due to the lack of spare parts provided by the U.S. in response to the erroneous sale. One of the F9F-8T aircraft would break Argentine speed records, before the aircraft ended service in 1971.

About 60 F9F Cougar models were converted to unmanned target drones with the designations (depending on the target configuration) of QF-9J, F9F-8K, and QF-9G, and flew in practice ranges such as China Lake target area through the 1960s.

An early prototype XF9F-6 swept-wing aircraft, BuNo 126671, flies ahead of a late-production straight-wing F9F-5 during a flight test in 1951. Many of the total of 1,385 straight-wing F9Fs that were built went on to serve in Korea. The Navy had always been interested in a swept-wing F9F aircraft, but swept-wing aerodynamics were a new and developing area of study, and accordingly, a straight-wing version of the airplane was developed first. (J.W. Hawkins Collection)

With its air brakes (speed brakes) fully deployed, F9F-9P Cougar, BuNo 141712, from the VFP-61 group, lands at Mitscher Field, in Miramar NAS in California in 1957. (John Hawkins Collection)

The first version of the Cougar, the F9F-6, is basically a redesigned version of the F9F Panther with both wings and tail section swept back. Mounted in the nose of the F9F-6 are the same four cannons as were carried in the nose of the Panther. Production of the F9F-6 began in mid-1952 and ended in mid-1954. (Grumman archives via Cradle of Aviation)

The F9F-6P is the photo reconnaissance version of the F9F-6 Cougar. This model required a slight extension of the nose section of about one foot in length in order to accommodate the various cameras that were fitted in the nose. Cameras included a nose-mounted, tri-metrogen K-17 camera and photo recorder gun camera. Production of this model began in 1954 and ended in 1955. (U.S. Navy)

F9F-6

Wingspan:	34 feet, 6 inches
Length:	41 feet, 5 inches
Height:	12 feet, 3 inches
Empty weight:	11,255 lbs
Gross weight:	18,450 lbs
Powerplant:	Pratt & Whitney J48-P-6A (first 30)/J48-P-8
Armament:	Four 20mm cannon and 2,000 lbs on two wing pylons
PERFORMANCE	
Maximum Speed:	654 mph
Service Ceiling:	44,600 feet
Range:	932 miles
Crew:	1
Number built:	646

F9F-6P

Wingspan:	34 feet, 6 inches
Length:	42 feet, 2 inches
Height:	12 feet, 3 inches
Empty weight:	11,255 lbs
Gross weight:	18,450 lbs
Powerplant:	Pratt & Whitney J48-P-8
Armament:	None
PERFORMANCE	
Maximum Speed:	654
Service Ceiling:	44,600 feet
Range:	932 miles
Crew:	1
Number built:	60

The F9F-7 model was developed in order to accommodate the new Allison J33-A-16A engine. Problems arose with the engine, however, and only the first 118 aircraft of this model had the Allison power plant, with the later aircraft changed back to the original Pratt & Whitney J48 engine, and earlier models retrofitted as well. (Ken Neubeck)

The F9F-8, the next major version of the Cougar, has a larger wing area than the previous versions, enabling it to carry more bombs or fuel in wing-mounted racks. Production of this model began in 1954 and ended in early 1957. This was one version of the aircraft that was used by the Blue Angels. (Ken Neubeck)

F9F-7

Wingspan:	34 feet, 6 inches
Length:	42 feet, 2 inches
Height:	12 feet, 3 inches
Empty weight:	11,225 lbs
Gross weight:	18,905 lbs
Powerplant:	One Allison J33-A-16A
Armament:	Four 20mm cannon and 2,000 lbs on two wing pylons
PERFORMANCE	
Maximum Speed:	628
Service Ceiling:	40,200 feet
Range:	1,157 miles
Crew:	1
Number built:	168

F9F-8

Wingspan:	34 feet, 6 inches
Length:	42 feet, 2 inches
Height:	12 feet, 3 inches
Empty weight:	11,866 lbs
Gross weight:	20,098 lbs
Powerplant:	One Pratt & Whitney J48-P-8A
Armament:	Four 20mm cannon and 2,000 lbs on four wing pylons
PERFORMANCE	
Maximum Speed:	647
Service Ceiling:	42,000 feet
Range:	1,208 miles (combat range)
Crew:	1
Number built:	601

The F9F-8P is the photo reconnaissance version of the F9F-8 Cougar. In contrast to the earlier F9F-6P model, this aircraft required a significant extension of the nose – about two extra feet – to accommodate 14 or more cameras in the nose section. Production of this model ran from 1955 to 1957. (Ken Neubeck)

The F9F-8T, the two-seat trainer version of the F9F-8 Cougar, features a significant extension of the fuselage in order to accommodate a second crew member. This model was also used by the Blue Angels, and it was the only F9F Cougar model to see action in Vietnam and to be sold overseas. (Ken Neubeck)

F9F-8P

Wingspan:	34 feet, 6 inches
Length:	44 feet, 9 inches
Height:	12 feet, 3 inches
Empty weight:	12,246 lbs
Gross weight:	18,421 lbs
Powerplant:	One Pratt & Whitney J48-P-8A turbojet
Armament:	None
PERFORMANCE	
Maximum Speed:	637 mph
Service Ceiling:	41,500 feet
Range:	960 miles
Crew:	1
Number built:	110

F9F-8T

Wingspan:	34 feet, 6 inches
Length:	44 feet, 5 inches
Height:	12 feet, 3 inches
Empty weight:	12,787 lbs
Gross weight:	20,600 lbs
Powerplant:	One Pratt & Whitney J48-P-8A turbojet
Armament:	Two 20mm cannon and 2,000 lbs on four wing pylons
PERFORMANCE	
Maximum Speed:	642 mph
Service Ceiling:	42,000 feet
Range:	1,208 miles (combat range)
Crew:	2
Number built:	399

Nose Variations

F9F-6, F9F-7, and F9F-8 Models
The nose section has four cannon holes that are located in lower section. (Ken Neubeck)

F9F-8P Photo Reconnaissance Model
The nose section has no cannons installed and is extended to house camera equipment. (Ken Neubeck)

F9F-8T Trainer
The nose section has only two cannon holes in the lower section. (Grumman archives via Cradle of Aviation)

Canopy Variations

F9F-6, F9F-7, and F9F-8 Models
The canopy allows seating for one pilot. (Grumman archives via Cradle of Aviation.

F9F-8P Photo Reconnaissance Model
The canopy allows seating for one pilot. (Grumman archives via Cradle of Aviation)

F9F-8T Trainer
The canopy has the instructor in front and student pilot in the rear. (Grumman archives via Cradle of Aviation)

With the top of its vertical stabilizer trimmed off, this F9F-6, BuNo 128870, is testing a modified tail section in October of 1952. (J.W. Hawkins collection)

Outfitted with a test probe and with bombs mounted on the two wing pylons, this F9F-6, BuNo 127219, is undergoing tests in April of 1954. (J.W. Hawkins collection)

The unique structure of the F9F-6 includes external fuel tanks mounted on pylons on the wings near the fuselage. (J.W. Hawkins collection)

The location of the canopy in relation to the fuselage and the small tail section are evident in this overhead view of an F9F-6 flying through the clouds. (J.W. Hawkins collection)

After 1955, the U.S. Navy converted the F9F Cougar to a light gray paint scheme in lieu of the dark blue paint scheme that was used in the early Cougar models. This gray scheme would be the principal paint scheme for the Cougar, as well as other U.S. Navy aircraft, for the remainder of its active service. (Grumman archives via Northrop Grumman History Center)

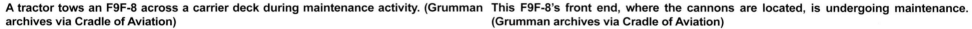

The final assembly line at the Grumman factory in Bethpage, New York, puts together F9F production aircraft. (Grumman archives via Cradle of Aviation)

The sixth production F9F-8 aircraft, BuNo 131083, undergoes flight tests over eastern Long Island in 1954. (Grumman archives via Cradle of Aviation)

A tractor tows an F9F-8 across a carrier deck during maintenance activity. (Grumman archives via Cradle of Aviation)

This F9F-8's front end, where the cannons are located, is undergoing maintenance. (Grumman archives via Cradle of Aviation)

This F9F-6 test aircraft has a miniature F9F-6 model mounted on the nose probe for collecting aerodynamic data. (J.W. Hawkins collection)

The F9F-8 wing was modified from the original F9F-6 design to be stronger in order to carry three pylons under each wing. (Grumman archives via Cradle of Aviation)

This F9F-8 features a smooth underbody with no underwing pylons. The wings on the F9F-8 were made stronger than those on previous models in order to increase wing station capabilities. (Grumman archives via Cradle of Aviation)

The underside of this F9F-8 aircraft shows the complement of six wing pylons; two for external fuel tanks and four for sidewinder missiles. (Grumman archives via Cradle of Aviation)

This F9F-7, BuNo 130763, once served on the USS *Randolph*. Later, after active service, it was used for firefighting practice at NAS Lakehurst in New Jersey. It has since been restored and is on display at the Cradle of Aviation Museum in Long Island. In 1953, 169 F9F-7 aircraft were produced. Initially, the F9F-7 incorporated an engine change. The new Allison J33-A-16A engine proved unreliable, however, and only 50 of the F9F-7 models were delivered with it. The remaining aircraft went back to the original J48 engine. (Ken Neubeck)

This frontal view of the F9F-7 shows the position of the four holes for the cannons and front view of the two inlet sections mounted on the fuselage. (Ken Neubeck)

All four of the holes for the cannons are oblong in shape to fit with aerodynamic flow that rides along the nose section. (Ken Neubeck)

There are no attachments on the nose of this F9F-7. Below the front gun ports is a rectangular assembly that contains the UHF homing antenna. Changes to the nose occurred in latter Cougar models. (Ken Neubeck)

As a result of some carrier landing incidents, the front nose section of later models, such as this F9F-8, had a barricade deflector attachment for engaging the ship's barrier when missed tail hook landings occur. (Ken Neubeck)

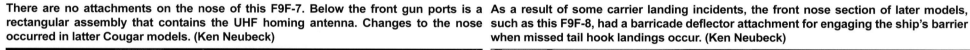

13

On display at the USS *Intrepid* Air and Space Museum in New York City, this F9F-8, BuNo 141117, originally served with the VF-111 and VF-94 squadrons and later became a chase aircraft. It has been restored to a black-and-yellow paint scheme to reflect the VF-51 Squadron that served aboard the *Intrepid* in 1956. Featuring a longer center fuselage than the F9F-6, the F9F-8 version came into production in 1953. Grumman built 601 of the F9F-8 aircraft for the Navy between 1953 and 1957. In 1962, the F9F-8 was redesignated the F-9J. (Ken Neubeck)

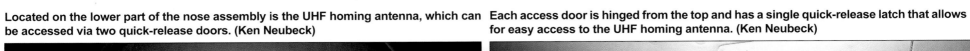

An in-flight refuelling probe is a feature of later F9F models such as this F9F-8. Tabs on the probe locked it in place during refueling. (Ken Neubeck)

Located below the base of the in-flight refueling probe is the barrier deflector on the lower fuselage, below the inside cannon exit ports. (Ken Neubeck)

Located on the lower part of the nose assembly is the UHF homing antenna, which can be accessed via two quick-release doors. (Ken Neubeck)

Each access door is hinged from the top and has a single quick-release latch that allows for easy access to the UHF homing antenna. (Ken Neubeck)

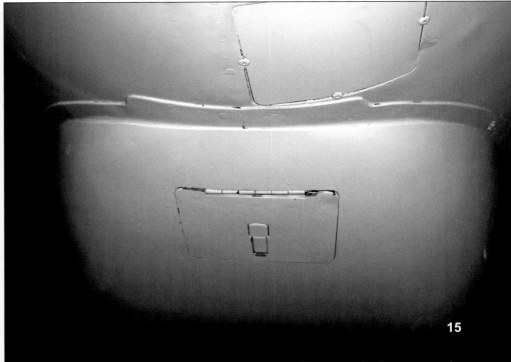

The F9F employs two main landing gears that pull inboard when retracted during flight, and a nose landing gear that pulls towards the rear during flight. (Ken Neubeck)

This rear view of the nose landing gear shows the strut and doors. Here the air brakes (speed brakes) are retracted and flush with the lower fuselage. (Ken Neubeck)

All of the major components of the nose landing gear are seen here from the side: the wheel, the fork section, shimmy damper, and scissors assembly. The nose landing gear design remained consistent for all of the different F9F models, including the two-seat trainer. (Ken Neubeck)

This forward view of the nose landing-gear assembly for the F9F Cougar shows the dual fork assembly that holds the wheel assembly. A set of two doors, one on each side of the nose landing gear, close over the gear when it is retracted during flight. The nose landing gear of this F9F is displayed here partially extended. The perforated air brakes located behind the landing gear are not deployed but flush with the lower fuselage of the aircraft. (Ken Neubeck)

On this particular F9F, the landing gear is fully extended. Also fully extended are the two perforated air brakes, located just behind the landing gear and hinged on the rear of the nose landing gear wheel well. The nose wheel steering unit sits on top of the fork, with the shimmy damper spring located on the left side of the strut assembly. There are two tow rings, one on either side of the front wheel, and there are two tie-down rings located on the upper part of the strut assembly. (Ken Neubeck)

The shimmy damper is connected to the top and bottom of the nose wheel steering unit and is also connected by link arms to the strut at the top. (Ken Neubeck)

Stencilled on the body of the F9F are maintenance instructions, including instructions regarding nose strut and tire pressure written on both the right and left doors of the nose landing gear. Maximum nose wire tire pressure is 175 PSI. (Ken Neubeck)

This left side view of the nose landing gear shows the shimmy damper spring that is located on the left side of the strut with a linkage on the top that ties to the nose wheel steering unit. There are six circular cutouts in the two-piece wheel assembly. The dual fork assembly is attached to the wheel by means of a perforated hex nut and cotter pin that retains the bearing assembly. An instruction for towing is printed on the fork of the nose wheel on this particular aircraft. (Ken Neubeck)

Each of the two door haves of the nose landing gear is connected with a linkage that goes to the mechanism located in the top of the wheel well. (Ken Neubeck)

In this view from the rear of the nose landing gear, with air brakes retracted into the fuselage, the scissor assembly is partially extended on the nose strut. (Ken Neubeck)

There is a bracket assembly located on top of the shimmy damper that is connected to the top of the nose wheel steering unit. The scissor assembly used to extend the nose landing gear is located in the rear of the strut and is fully extended on this aircraft. A pushrod linkage on the on the left side aids in the full extension of the scissors assembly. Pneumatic pressure for the strut in the fully extended position is in the range of 215 to 230 PSI. (Ken Neubeck)

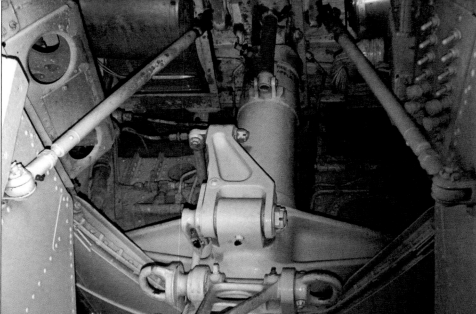

The nose landing gear strut goes to the top of the wheel well into a bracket assembly that also has the linkages to the doors attached to it. (Ken Neubeck)

Located on the left side above the nose gear door and inside the wheel well, are electrical harnessing, alternators, circuit breakers, and an inverter power distribution box. (Ken Neubeck)

The nose landing gear scissor assembly, seen here from the rear, pivots on a rod that is retained by cotter-pin/hex-nut assemblies on the bottom and middle of the gear scissor assembly. On this particular aircraft, the strut is partially retracted and the linkage mechanism, which includes pushrods and a slotted bracket on the left of the strut, is pulled up. The tie-down rings that are located on the upper part of the strut assembly, at the top portion of the scissor assembly, can rotate as needed. (Ken Neubeck)

The relays seen at left are among the number of electrical components that are mounted on the ceiling of the nose landing gear wheel well. (Ken Neubeck)

The cutout in the rear wall of the nose gear wheel well provides access to some of the interior electrical and hydraulic lines. (Ken Neubeck)

The nose landing gear wheel well has both electrical and hydraulic components located inside the well area, with a number of electrical boxes that are located on the two side shelves and on the ceiling. The large green cylinder mounted on the ceiling provides emergency air for deployment of the landing gear and dive brake. (Ken Neubeck)

On display at the Pima Air and Space Museum in Tucson, Arizona, is this F9F-8P, BuNo 144425. In all, 110 photo-reconnaissance F9F-8P Cougars were built, beginning in 1955. A single-seat aircraft, the F9F-8P had had an extended nose that housed reconnaissance cameras. This aircraft was taken out of service after only two years. (Ken Neubeck)

The F9F-8P model has a camera mount rather than a barricade deflector on the lower nose. Most F9F-8P aircraft were fitted wth the in-flight refueling pod. (Ken Neubeck)

A single CA13B Camera with a CIL-8 cone fits against the inside of the forward camera window on the nose of the F9F-8P. (Ken Neubeck)

Located on both sides of the extended nose of the F9F-8P aircraft are two windows where the two CA-17 cameras are positioned. (Ken Neubeck)

The large access panel that contains the windows is hinged on the top and is used during ground maintenance of the CA17 and CA-13B cameras. (Ken Neubeck)

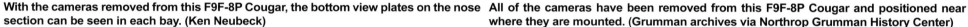

With the cameras removed from this F9F-8P Cougar, the bottom view plates on the nose section can be seen in each bay. (Ken Neubeck)

All of the cameras have been removed from this F9F-8P Cougar and positioned near where they are mounted. (Grumman archives via Northrop Grumman History Center)

The full complement of all the cameras used in the F9F-8P Cougar includes the CA-17 and CA-13B cameras, which along with cone attachments, are displayed outside of this aircraft. (Grumman archives via Northrop Grumman history Center)

A maintenance worker replaces one of the many cameras that are located in the front bay of the forward fuselage of the F9F-8P Cougar. (Grumman archives via Northrop Grumman History Center)

The F9F-8P Cougar is a photo reconnaissance version that is a further enhancement of the earlier F9F-6P aircraft. Production of this model began in 1955. (Grumman archives via Cradle of Aviation)

In order to accommodate 14 cameras into the nose, the length of the F9F-8P was extended by an additional two feet from the F9F-8 model. (Grumman archives via Cradle of Aviation)

All of the production F9F-8P Cougar models were painted in the light gray and white paint scheme that was used by the U.S. Navy at the time. (Grumman archives via Cradle of Aviation)

The F9F-8P is the longest version of the F9F Cougar. Its extended nose design brought the overall length of the aircraft to more than 44 feet. (Grumman archives via Cradle of Aviation)

Orange trim identifies this TF-9J, BuNo 147397, as a trainer. The aircraft is on display at the Pima Air and Space Museum in Tucson, Arizona. Originally designated the F9F-8T, the two-seat trainer version of the Cougar was introduced in 1955 and a total of 399 of them were manufactured. In 1962, the F9F-8T aircraft were redesignated as TF-9J, and in 1966 they were used to direct airstrikes in South Vietnam, making them the only version of the Cougar to see combat action. (Ken Neubeck)

Because of the aircraft's specific mission, the nose design of the F9F-8T trainer differs significantly from that of previous F9F models. (Ben Peck)

The four cannons in the nose have been reduced to two cannons in the trainer in order to accommodate additional equipment. (Ben Peck)

The trainer version also lacks the bulge in the lower nose that housed the search radar scanner in previous F9F models. (Ken Neubeck)

In order to accommodate the extra seat that was added to the tandem seat design of the F9F-8T aircraft, the center fuselage had to be lengthened. (Ken Neubeck)

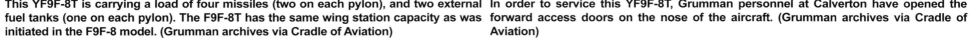

YF9F-8T, BuNo141667, a two-seat prototype with orange and white training markings, is on the ground at Grumman's Calverton facility in 1955. (Grumman archives via Cradle of Aviation)

Outfitted with a pitot test tube installed on its nose, YF9F-8T, BuNo 141667, conducts flight tests over the eastern Long Island facility. (Grumman archives via Cradle of Aviation)

This YF9F-8T is carrying a load of four missiles (two on each pylon), and two external fuel tanks (one on each pylon). The F9F-8T has the same wing station capacity as was initiated in the F9F-8 model. (Grumman archives via Cradle of Aviation)

In order to service this YF9F-8T, Grumman personnel at Calverton have opened the forward access doors on the nose of the aircraft. (Grumman archives via Cradle of Aviation)

The F9F-8T was designed to provide the most realistic setting possible for the pilot under training in the rear section. A small canopy section, low enough to fit under the regular canopy when the aircraft was in flight, hightened the realistic setting of the student pilot's station. Many of the F9F Cougar's maintenance instructions were written on the appropriate locations of the aircraft as can be seen on this example. In the background is another Grumman carrier-based aircraft, the F11F Tiger, which was originally designated the F9F-9. (Grumman archives via Northrop Grumman History Center)

An early production F9F-8T complete with test probe appears at an air show at Long Island MacArthur airport in November of 1957. (Ray Neubeck)

This early production F9F-8T features orange training markings on its nose and the trailing edges of wings and tail. (Grumman archives via Cradle of Aviation)

With the canopy pulled back and the two pilots exposed to the air, this later-production F9F-8T flies over eastern Long Island. (Grumman archives via Cradle of Aviation)

The same production F9F-8T aircraft is flying with the canopy closed. This aircraft does not have the training markings yet. (Grumman archives via Cradle of Aviation)

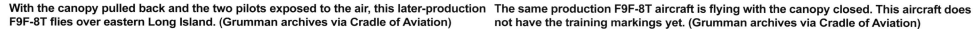

The front nose section of the F9F-6, -7 and -8 models holds the four 20mm M-3 cannons and their ammunition drums. (Ken Neubeck)

The front glass section of the canopy is made of bulletproof glass. This F9F-6, BuNo 128109, is on display at the NAS museum in Pensacola, Florida. (Ken Neubeck)

The front glass section of the canopy is joined to a left and right section by means of a bracket. High-speed aircraft such as the F9F required smooth surfaces on the fuselage for top performance, hence there is a note regarding lacquer coating on the top of the fuselage. (Ken Neubeck)

Removal of the nose section reveals many components. The barrels of the right outboard and inboard 20mm cannons can be seen in the lower part of the nose, with the ammunition tanks and chutes located at top, center. The rounded assembly at the front is the D/F loop antenna, with the D/F transmitter and receiver box located directly behind it. Behind the equipment and in front of the cockpit is an armored bulkhead that provides protection to the pilot. (Grumman archives via Northrop Grumman History Center)

The canopy for the F9F single-seat aircraft consists of two sections: a fixed front section and a moveable rear section. (Ken Neubeck)

The canopy's fixed front section consists of three parts: a front piece and two side pieces that are held by brackets and cover the front console area. (Ken Neubeck)

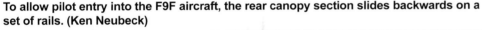

To allow pilot entry into the F9F aircraft, the rear canopy section slides backwards on a set of rails. (Ken Neubeck)

When the rear canopy section is fully opened, it slides back, past the headrest of the ejection seat. There is a small support bracket inside the canopy glass. (Ken Neubeck)

The left foot cutout is located at the lower left point of the insignia and it is spring-loaded at the bottom of the panel. (Ken Neubeck)

The left hand grip cutout is located in the top right portion of the insignia, just below the canopy. The bottom of the grip cutout panel is spring loaded. (Ken Neubeck)

In order to board an early F9F Cougar, such as this F9F-7, the pilot would climb into the left side of the aircraft by using the spring-loaded cutout panels that were built into the fuselage. There are two cutouts located on the left side for the left foot and left hand, along with two cutouts that are located on the right side for the right foot and right hand. The short boarding ladder located in front of the engine intake area is seen here in the retracted position. (Ken Neubeck)

A short ladder step for the left foot pulls out of the lower fuselage. The right foot then goes into the cutout slot that is located inside the engine intake splitter plate. (Ken Neubeck)

The left foot cutout panel is located in the stripe of the insignia. There are thus three steps on F9F-8 models, as compared with four steps on the earlier model. (Ken Neubeck)

On later F9F models, such as the F9F-8P aircraft, the boarding scheme was changed significantly. Entry is still accomplished from the left side of the aircraft but the steps have been arranged to accommodate the engine intake plate. The boarding ladder, located just below the left engine intake splitter plate, is shown in the extended position. A spring-loaded cutout panel slot for the right foot, located inside the engine intake area, is intended for the pilot's next step up. Another foot cutout slot for the left foot is located just below the canopy. In addition, black lines are painted on the fuselage as a visual guide for the pilot during boarding. (Ken Neubeck)

Located to the right of the target span assembly is the standby compass. The fuel level warning lights below the compass are missing labels. (Ken Neubeck)

On the cockpit of this F9F-7 aircraft, the fire warning circuit test and low-fuel level warning lights are labeled in red. Some modifications were made in the type of switches and lamp used in this model. (Ken Neubeck)

This F9F-6 cockpit has a sight unit and target span scale assembly located at the top of the cockpit panel and marked with the "No hand hold" warning. There are three rows of aircraft instrument gages on the main cockpit panel with the control stick located in front. The stick has a trigger button for firing the 20mm cannons. The cockpit remained consistent for the single-seat F9F models and for the pilot's position aboard the two-seat F9F-8T trainer version. (Ken Neubeck)

The gage with aircraft outline is the G-2 remote compass indicator and the gage to the left of it is the airspeed and Mach number indicator. (Ken Neubeck)

The sight unit, consisting of a glass projection screen and a dial indicator, is mounted on a base assembly that is attached to the top of the cockpit panel. (Ken Neubeck)

The red cover to the far right of the cockpit conceals the tail skid control switch. There are several circuit breakers located on the inside of the right console. (Ken Neubeck)

The control stick is mounted to the floor. A dual set of rudder pedals with linkages is located underneath the cockpit panel. (Ken Neubeck)

The main instrument panel for the forward station of the F9F-8 single-seat model and the F9F-8T trainer incorporated some minor variations from the instrument panel on the F9F-7 and F9F-6. On the F9F-8 and -8T, the individual instruments were mounted with flanges on the panel front. On the earlier designs of the aircraft, the instruments were rear mounted. The fire warning light and fire warning light test switch were relocated from the previous positions at the top right side of the panel to the panel's top center, just below the gun sight. The aircraft's BuNo, in this case BuNo 142466, was mounted on a plate at the top left of the panel. (Grumman archives via Northrop Grumman History Center)

The main instrument panel in the aft pilot's position aboard the F9F-8T trainer differed somewhat from the panel in the forward pilot position. There is no gun sight on the top of the panel, nor is there a Low Altitude Bombing System (LABS) control box located on the top left of the panel. The aft position has a small canopy section to simulate the forward position. The individual flight instruments on the panel were identical to those used by the forward pilot and most were located in the same position. One difference was the use of a clock in place of the G-2 remote compass on the lower right side of the panel. A shroud is provided for simulated night flight. (Grumman archives via Northrop Grumman History Center)

Located on the left side of the single-seat F9F cockpit is the handle for opening the rear canopy section. Emergency handles in the cockpit are painted red. (Ken Neubeck)

The red handle on the left is the emergency manual landing gear handle. The red handle to the right is for the emergency air start igniter. (Ken Neubeck)

The left console contains the throttle controls and flight control panels. The lever on the far right is the landing gear control lever. (Ken Neubeck)

On the right console are the control panels for the radios and interior and exterior lighting. The panel on the right is for the combat hydraulic system power. (Ken Neubeck)

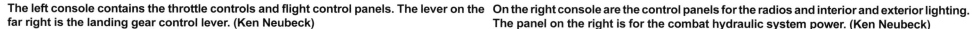

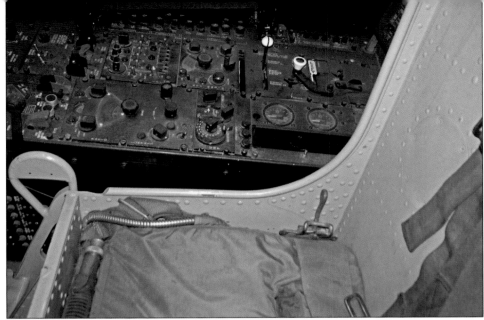

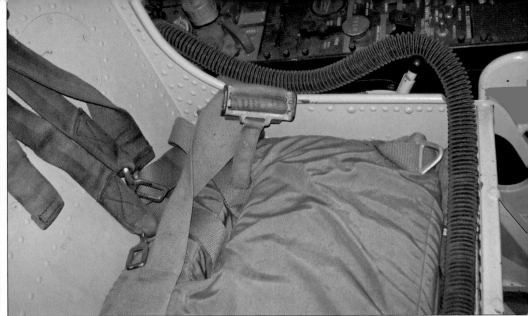

Located next to the right knee brace is the weapons control panel with protected red firing switches. Two hydraulic gages are further to the right. (Ken Neubeck)

The original ejection seat that was used in the F9F-6, F9F-7, and some F9F-8 models was designed by Grumman. Located just above the headrest is the handle for pulling down a cover to protect the pilot's face during ejection. (Ken Neubeck)

Left of the seat is the red handle for the wheel brakes emergency control and red button for emergency longitudinal control. Seat belts and harnesses are standard. (Ken Neubeck)

The headrest is attached to the ejection seat by brackets. The Grumman ejection seat was replaced by the Martin Baker Mark V seat on the F9F-8T. (Ken Neubeck)

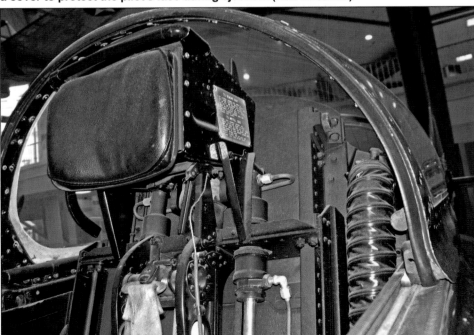

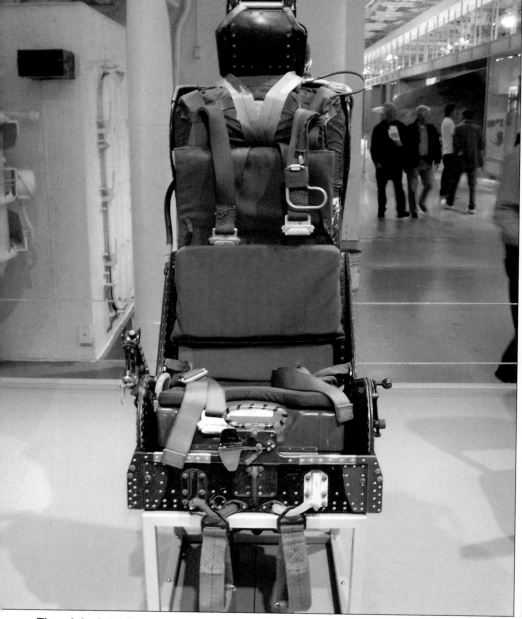

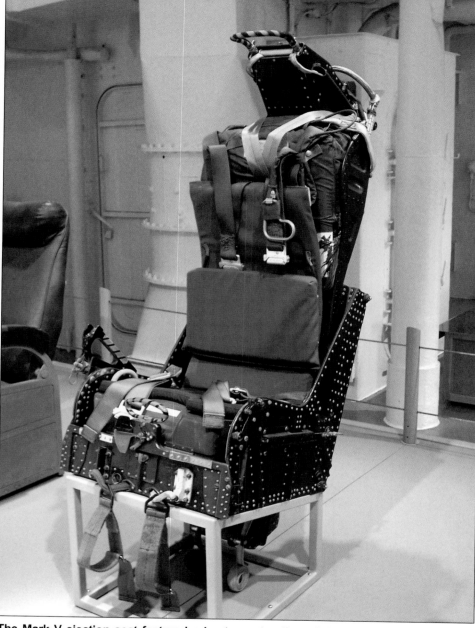

The original ejection seat used in the F9F aircraft was made by Grumman. In 1957, the Navy successfully conducted tests in Patuxent River, Maryland, with a modified F9F-8T using a Martin Baker Mark V ejection seat situated in the rear position. A successful ejection was made with the aircraft on the runway. As a result, the Mark V was incorporated into most of the F9F-8T trainer aircraft and in some of the earlier single-seat models. It would be used on several other U.S. Navy aircraft as well. (Ken Neubeck)

The Mark V ejection seat featured robust construction of the seat design in order to handle ground ejections along with other emergency features. There is a crotch pull loop located in front of the seat as well as a loop located on top of the head rest that is used for the face curtain. A set of lower leg restraints are attached to the lower part of the seat buckets. The seat bucket is black metal with rows of rivets. Padding for the seat cushion is in olive drab. (Ken Neubeck)

Seat cushions and pads sit inside a seat frame that is constructed with many sets of rivets for rigid construction. (Ken Neubeck)

Located beneath the lap belts and seat cushion is the survival kit container. On the right side of the crew member is the manual override level for belt release. (Ken Neubeck)

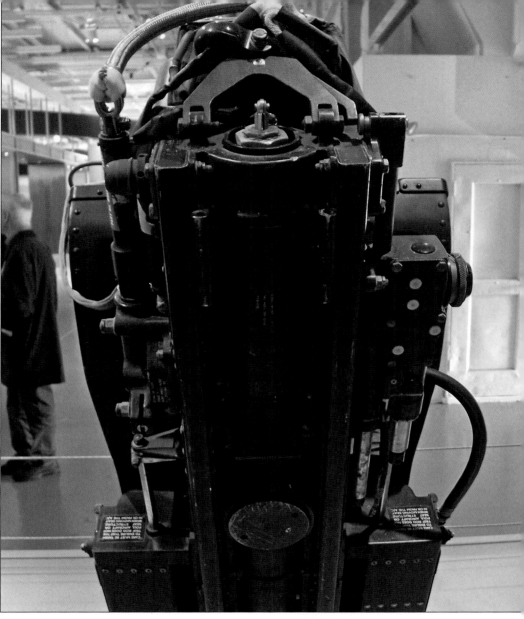

On the rear of the ejection seat are the seat rails that ride on the aircraft rails during ejection from the aircraft. The top of the seat is capable of breaking the glass of the canopy during ejection. The mechanism located to the left of the rails is the drogue gun, which fires about two seconds after the seat clears the aircraft. The mechanism located on the right side of the rails is the time-release mechanism that provides the necessary time delay for the ejection after the initial seat movement, which starts slowly. (Ken Neubeck)

Each air intake port for the basic F9F model is located at mid-fuselage and is flared with the wing. (Ken Neubeck)

Due to the extended fuselage on the two-seat F9F-8T trainer, a splitter plate was added to the front of the intakes for improved intake airflow. (Ken Neubeck)

The extended fuselage of the F9F-8P meant that a splitter plate had to be installed in front of the intake for better aircraft stability during flight. (Ken Neubeck)

The right side intake is marked the same as the left, with warning text inscribed in the red outline of the entrance. (Ken Neubeck)

Later F9F-8 models have the engine splitter plate, along with some variations on the red warning markings from previous models. (Ken Neubeck)

Directly below the engine splitter plate and engine intake are the air brake panels, which on this particular F9F-8T aircraft are closed. (Ken Neubeck)

Air brakes on this aircraft are in the closed position and the panel is flush with the lower fuselage of the aircraft. (Ken Neubeck)

As indicated by maintenance instructions on the outside of the air brake, there are vent valves for certain components that use bottled air for emergencies. (Ken Neubeck)

The combustion chambers of the engine are seen here at left, while the exhaust section is to the right. The engine used in the F9F Cougar was the J48-P-8A Pratt & Whitney turbojet, which was a licensed version of the Rolls Royce RB.44 Tay engine. The J48-P-8A turbojet was originally used in the F9F-5 Panther fighter and then incorporated into most F9F Cougar models, except for original shipments of the F9F-7 that used an Allison J33-A-16 turbojet instead. (Ken Neubeck)

46

Highlighted in this close up are the compressor blades, which are located in the intake section of the engine. (Ken Neubeck)

Located immediately behind the nine individual combustion chambers is the exhaust nozzle section of the engine. (Ken Neubeck)

The J48 engine fits into the Cougar body to interface with the fuel lines and components mounted on the rear bulkhead area in the forward fuselage. (Ken Neubeck)

This cutaway view of the front portion of the engine shows the two-stage intake fan section and two of the combustion chambers outside the engine. (Ken Neubeck)

On display at the Pima Air and Space Museum, this restored F9F-8 Cougar, BuNo141121, is painted in the orange and blue colors of the VT-23 training squadron, better known as *The Professionals* and based out of Kingsville, Texas, at the time. The F9F-8 model was the last major significant single-seat version of the Cougar, with over 601 aircraft produced. The first version of this model flew in 1953. (Ken Neubeck)

This view of the right wing of the F9F-6 shows the wing flaring from the engine intake section. (Ken Neubeck)

Located outboard of the window is a fence assembly that controls airflow. It rides on the top of each wing and wraps around the forward edge of the wing. (Ken Neubeck)

All of the trailing edge flaps on this F9F-6 are in the closed position. The wing is painted in navy blue, with white wing tips and silver leading edges. (Ken Neubeck)

In order to allow them to accommodate six underwing stations, the wings of later Cougar models, like this F9F-8, were made larger than those of the F9F-6. (Ken Neubeck)

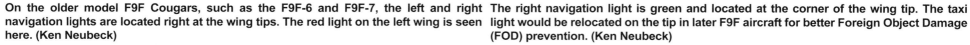

Located on the outboard side of the engine intake on the left wing only, is the landing/taxi lamp assembly. (Ken Neubeck)

The landing light is located just outboard of the wingfold area on the left side of the aircraft. (Ken Neubeck)

On the older model F9F Cougars, such as the F9F-6 and F9F-7, the left and right navigation lights are located right at the wing tips. The red light on the left wing is seen here. (Ken Neubeck)

The right navigation light is green and located at the corner of the wing tip. The taxi light would be relocated on the tip in later F9F aircraft for better Foreign Object Damage (FOD) prevention. (Ken Neubeck)

The F9F-8 design incorporated changes in the wing design that strengthened it to allow it to hold an additional wing pylon station. (Ken Neubeck)

Along with the new wing design on the F9F-8, the wing navigation lights were also set further to the rear. (Ken Neubeck)

The wing design on the F9F-8T trainer aircraft was identical to the wing design of the F9F-8. (Ken Neubeck)

Relocating the navigation lights back from the leading edge of the wing allowed for better FOD protection along with the wing tip protrusion. (Ken Neubeck)

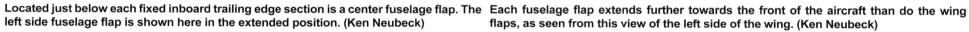

This view of the left wing of this F9F-8P aircraft shows the location of the fence section at mid-wing. The paint scheme is light gray with red wing tips. (Ken Neubeck)

This view from the rear of the left wing shows the flap extended, exposing the fixed inboard trailing edge section on the fuselage. (Ken Neubeck)

Located just below each fixed inboard trailing edge section is a center fuselage flap. The left side fuselage flap is shown here in the extended position. (Ken Neubeck)

Each fuselage flap extends further towards the front of the aircraft than do the wing flaps, as seen from this view of the left side of the wing. (Ken Neubeck)

Their wings folded up, two F9F-8T aircraft sit side by side on the deck of an aircraft carrier. (Grumman archives via Cradle of Aviation)

The wings are folded up and the cockpit is open on this early-production F9F-8T parked on a runway. (Grumman archives via Cradle of Aviation)

Two maintenance men perform service on this particular F9F-8T trainer aircraft during the late 1950s. When F9F aircraft parked on the decks of aircraft carriers, hydraulic actuators in the aircraft wings would be used to fold up the wings. This practice allowed aircraft to be placed in areas where they could be worked on without taking up valuable real estate on the carrier deck. (Grumman archives via Cradle of Aviation)

On display at the USS *Lexington* Museum On The Bay in Corpus Christi, Texas, is an F9F-8T Cougar with its wings folded upward. (Mark M. Hancock)

The wings can be folded at different angles, but 90 degrees to the plane of the wing, as shown in this museum F9F aircraft model, is typical. (Ken Neubeck)

This wing fold on this F9F-8 aircraft is positioned at more than 90 degrees from the plane of the wings and at the maximum folding angle inwards for the F9F aircraft. By having the fold point at the outside of the engine inlet, the footprint of the width allows for convenient storage on a busy aircraft-carrier deck. (Ben Peck)

The wing folds at the point of the outboard panel of the main landing gear and near the weapons pylon. (Grumman archives via Northrop Grumman History Center)

The wing-fold area includes links and hydraulic components. The pylon is here fitted with sway braces for an external fuel tank. (Grumman archives via Northrop Grumman History Center)

The F9F Cougar's ability to fold its wings allowed not only for better space management on carrier decks but also, as seen here, made it easier to tow the aircraft through tight areas at airports located on land. The tow bar from the tractor connects to tow points that are located on the outside of the nose landing gear wheel assembly. (Grumman archives via Northrop Grumman History Center)

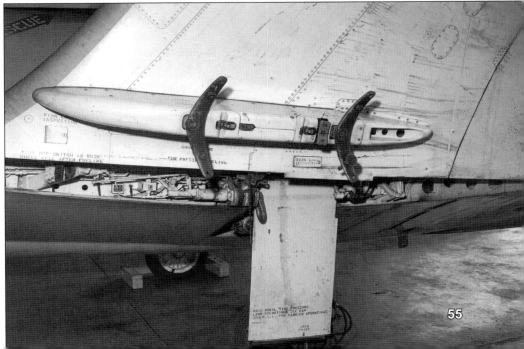

This view of the rear portion of the F9F shows the position of the landing gear, wing, and tail section with respect to each other. (Ken Neubeck)

As seen in this inside view of the left main landing gear wheel, 12 bolts fix the inside half of the wheel construction to the outside half. (Ken Neubeck)

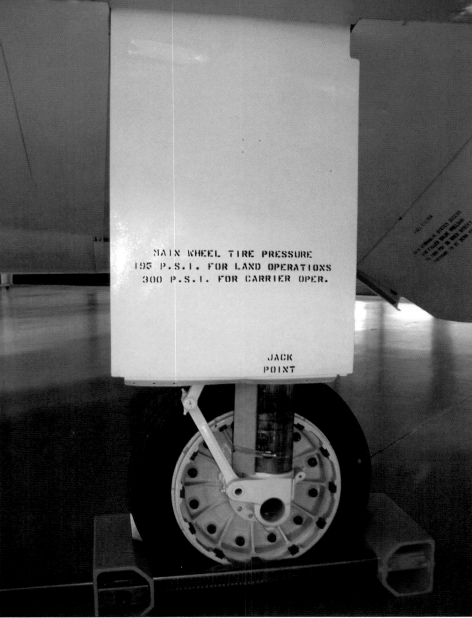

There is a rectangular panel located on the outboard side of each main landing strut assembly on the F9F Cougar. The scissors assembly is located on the forward side of the strut and is fully extended when on the ground. Maintenance instructions as to tire air pressure and jack point for the F9F Cougar landing gear are printed on the outboard panel. Air pressure for the tires varies depending on location – 195 psi for land-based operation and 300 psi for carrier-based operations. (Ken Neubeck)

The right main landing gear assembly of the F9F Cougar aircraft consists of a wheel assembly that is mounted on a bearing assembly on the inboard side of the strut. The strut is linked to the aircraft by means of a large hydraulic actuator that pulls the landing gear into the fuselage. The tires used for the main landing gear are different from those on the nose landing gear and are maintained at a higher air pressure. The landing gear design would remain consistent throughout the different F9F models. (Ken Neubeck)

The F9F Cougar's left main landing gear assembly mirrors the right main landing gear. The outboard panel that is mounted on the strut will become flush with the fuselage when the landing gear is retracted. The outboard panel is contoured at the end to fit with the unique contour of the fuselage. The landing gear strut, including the scissor assembly mounted on its front, is in the normal extended position when the aircraft is on the ground. (Ken Neubeck)

JACK POINT

This outboard close-up view of the left main landing wheel assembly shows some of the 12 bolts that hold the two halves of the wheel together. An additional 12 clips secure the rim of the wheel to the wheel assembly. The scissor assembly is mounted on a machined portion of the base of the landing gear strut. This base point has a lubrication port for the bearing on which the lower part of the scissor rotates. (Ken Neubeck)

This inboard view of the right main landing gear assembly on the F9F Cougar shows the strut assembly that is positioned towards the front, inside the outboard panel. The main landing gear actuator, located behind the top portion of the strut, helps retract the landing gear inboard after takeoff. At the center of the wheel assembly, the two halves of which are held together by 12 bolts, is the wheel bearing. (Ken Neubeck)

This view of the right main landing gear shows the inside of the wing cutout panel that is attached to the main landing gear strut. The main landing gear strut is positioned towards the rear of the cutout panel. An extension from the top of the strut bracket area serves as an attachment fixture to the hydraulic actuator that extends from the wheel well. Additional bracket support connects the lower part of the panel to the strut. (Ken Neubeck)

Visible in this inside view of the left main landing gear strut are two lines that are located on the front and back of the strut and that go to access ports. Located on the top inside of each main landing gear strut is a tie-down ring for securing the aircraft on the carrier. A peg on the lower rear of the strut engages a hook assembly that is located on the wall of the main landing gear wheel well. The inside of each panel is perforated. (Ken Neubeck)

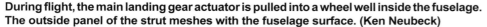

The top of the main landing gear strut has hydraulic lines that ride on the outside. The top portion of the strut extends towards the front of the aircraft. (Ken Neubeck)

The main landing gear actuator is mounted from inside the wing and hinges on the forward extruded section that extends from the landing gear strut. (Ken Neubeck)

During flight, the main landing gear actuator is pulled into a wheel well inside the fuselage. The outside panel of the strut meshes with the fuselage surface. (Ken Neubeck)

Mounted on one of the inside walls of the main landing gear wheel well is a hook assembly that engages with a peg located on the landing gear strut. (Ken Neubeck)

The F9F-8B was the same as the F9F-8, but with provisions for a low-altitude bombing system (LABS). (Grumman archives via Northrop Grumman History Center)

Changes to the F9F-8B from the F9F-8 were primarily internal and mainly affected some cockpit instrumentation. (Grumman archives via Northrop Grumman History Center)

In NATC markings, F9F-8B, BuNo 141140, prepares for a catapult launch from a carrier deck. (Grumman archives via Northrop Grumman History Center)

On display at NAS Jacksonville, Florida, F9F-8B, BuNo 131230, is painted in the markings of VF-81 group. (Gary Chambers)

When the rear section of the canopy is pulled all the way back, the clearance allows the pilot to enter the seat in the cockpit. (Ken Neubeck)

When the canopy is slid all the way back, its rear section hangs over a portion of the spine of the aircraft. (Ken Neubeck)

The view from the rear of the aircraft shows the rear canopy section and the spine area of the aircraft as it leads to the tail section. (Ken Neubeck)

As the spine area approaches the tail section, there is a slight contour that blends into the vertical stabilizer section of the tail. (Ken Neubeck)

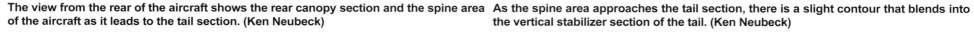

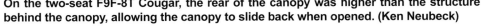

The area behind the rear of the canopy on the single-seat F9F Cougar is different from that on the two-seat version. (Ken Neubeck)

Since the canopy on the F9F single-seat model slides back to open, there could be no structures behind the canopy, other than the red anti-collision light. (Ken Neubeck)

On the two-seat F9F-8T Cougar, the rear of the canopy was higher than the structure behind the canopy, allowing the canopy to slide back when opened. (Ken Neubeck)

The red anti-collision light is nearer the rear pilot position on the F9F-8T than it is on the F9F. There are additional access panels behind the F9F-8T canopy. (Ken Neubeck)

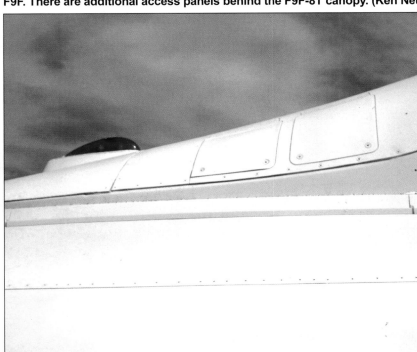

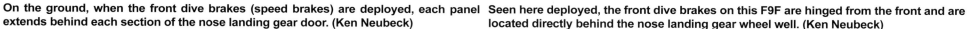
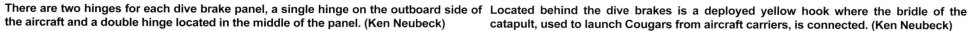

On the ground, when the front dive brakes (speed brakes) are deployed, each panel extends behind each section of the nose landing gear door. (Ken Neubeck)

Seen here deployed, the front dive brakes on this F9F are hinged from the front and are located directly behind the nose landing gear wheel well. (Ken Neubeck)

There are two hinges for each dive brake panel, a single hinge on the outboard side of the aircraft and a double hinge located in the middle of the panel. (Ken Neubeck)

Located behind the dive brakes is a deployed yellow hook where the bridle of the catapult, used to launch Cougars from aircraft carriers, is connected. (Ken Neubeck)

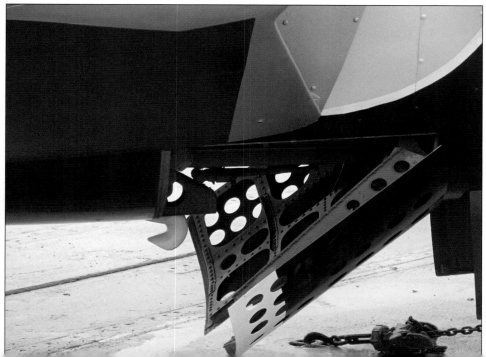

Each hydraulic actuator is connected to a frame section inside the lower fuselage and connected with hydraulic lines from the aircraft. (Ken Neubeck)

Each air brake actuator is connected to one hydraulic line. These lines link to a directional valve that is fed from aircraft hydraulic lines. (Ken Neubeck)

Located in the bay that houses the air brake actuators are four air vent valves used for servicing emergency air for landing gear, air brakes, and canopy. (Ken Neubeck)

Components that are located behind the air brakes are the two main landing gear and the lower flap control surfaces on the wing. (Ken Neubeck)

The F9F Cougar has an inboard wing section that flares towards the rear of the fuselage rather than a perpendicular wing style. (Ken Neubeck)

The flap on each wing is contoured with the wing up to the point where the inboard wing section begins. (Ken Neubeck)

The inboard section of each wing meets to the rear of the aircraft on either side of the exhaust section, located below the protruding vertical fin section. The overall pattern of the F9F Cougar's wing is similar to a dihedral or diamond shape, giving the aircraft a very distinctive look for the time period. The tail hook assembly is located in a small structural area in the lower fuselage, from which the tail skid assembly also protrudes. (Ken Neubeck)

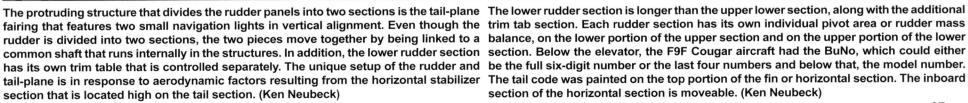

The protruding structure that divides the rudder panels into two sections is the tail-plane fairing that features two small navigation lights in vertical alignment. Even though the rudder is divided into two sections, the two pieces move together by being linked to a common shaft that runs internally in the structures. In addition, the lower rudder section has its own trim table that is controlled separately. The unique setup of the rudder and tail-plane is in response to aerodynamic factors resulting from the horizontal stabilizer section that is located high on the tail section. (Ken Neubeck)

The lower rudder section is longer than the upper lower section, along with the additional trim tab section. Each rudder section has its own individual pivot area or rudder mass balance, on the lower portion of the upper section and on the upper portion of the lower section. Below the elevator, the F9F Cougar aircraft had the BuNo, which could either be the full six-digit number or the last four numbers and below that, the model number. The tail code was painted on the top portion of the fin or horizontal section. The inboard section of the horizontal section is moveable. (Ken Neubeck)

The tail of F9F-7, BuNo 130763, is painted with the tail code letter R, which is the letter used by Carrier Air Group 17. The 126 marking is from the Navy Attack Squadron (VA-126) which used F9F aircraft for fleet instrument training during the late 1950s and thus has orange markings, indicative of a training squadron. This particular aircraft on display at the Cradle of Aviation museum did not actually serve in this squadron but has been painted in the training colors of VA-126 as part of its restoration. (Ken Neubeck)

The tail code NATC appears on this F9F-8 aircraft, BuNo 144276. The letters NATC are used by the Naval Air Test Center that is based out of Patuxent River in Maryland. This aircraft was originally assigned to the VA-44 squadron that was based out of NAS in Oceania. After retirement, the aircraft went on display at the Metro Richmond Visitor Center in Virginia before being transferred to the Patuxent River Naval Air Museum where it was changed to NATC markings. (Ken Neubeck)

This F9F-8T tail section from BuNo147397 is painted in the orange markings of the VT-10 training squadron that was based out of Pensacola, Florida. The F9F-8T was one of several naval aircraft that VT-10 squadron used for training purposes. VT-10 squadron specialized in aircraft that had a rear seat navigator position. The VT-10 was established in Pensacola in 1960. (Ken Neubeck)

The tail of this F9F-8P, BuNo144426, has tail code PP, which represents VFP-61, the reconnaissance group that was based out of NAS Miramar in California. The rear portion of the vertical fin section is painted in alternating red and white stripes. The BuNo and aircraft type are painted in small characters beneath the horizontal stabilizer. The tail skid is in the extended position on the lower fuselage. (Ken Neubeck)

Each elevator section is mounted on an oval plate that is flared to the rear of the tail. (Ken Neubeck)

The front and rear portions of the elevator assembly are moveable flight surfaces, with a shorter section spring tab located behind them. (Ken Neubeck)

The tail of this restored F9F-8P aircraft, BuNo 141675, has tail code PP, which represents VFP-61, the reconnaissance group that was based out of NAS Miramar in California. The major features of the tail section can be seen, with the tail skid and the black-and-white arresting hook assembly protruding from the lower part of the tail, along with the horizontal stabilizer assembly that is located at the middle of the vertical fin section. (Ken Neubeck)

Near the top of the bottom panel of the rudder section is a cutout where this panel pivots during flight. (Ken Neubeck)

The smaller trim tab section on the moveable elevator assembly is controlled by a pushrod that is located in the middle of the small flap. (Ken Neubeck)

In lieu of a single rudder flap section, there are two distinct sections that comprise the moveable rudder assembly of the F9F Cougar. The lower section is longer than the upper section and the two sections are divided by the fuselage protrusion that contains small tail landing lights. Both the lower and the upper sections pivot on an attachment point that connects to the fixed front vertical stabilizer section. (Ken Neubeck)

The lower portion of the tail assembly includes a tail skid assembly on the bottom and an arresting hook assembly extending from the rear. (Ken Neubeck)

The arresting hook is in the retracted position here, with associated linkages visible. There is a UHF blade antenna located on the right. (Ken Neubeck)

The arresting hook assembly is typically painted black and white. It is attached to linkage that attaches to a hydraulic actuator that deploys the hook. (Ken Neubeck)

Located before the arresting hook assembly is a hydraulically-driven tail skid assembly. An arresting hook release handle is located above the skid. (Ken Neubeck)

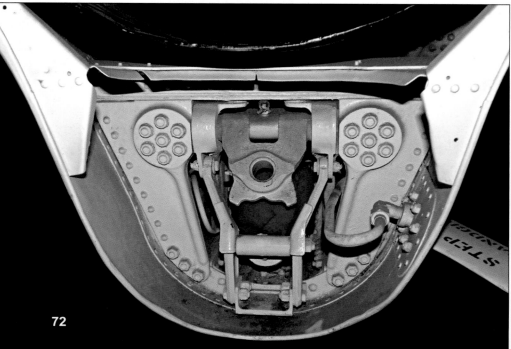

The engine exhaust area is located below the extended tail structure and above the tail skid and tail hook assembly. (Ken Neubeck)

Below and outside of the exhaust area is the tail skid, to the rear of which is a small blade antenna. (Ken Neubeck)

Located on both the left and right side of the engine exhaust area is a static vent port that is to be kept clear of any objects. (Ken Neubeck)

This is the exhaust area of the engine with the engine removed. Located in front of the engine are various fuel system components and fuel lines. (Ken Neubeck)

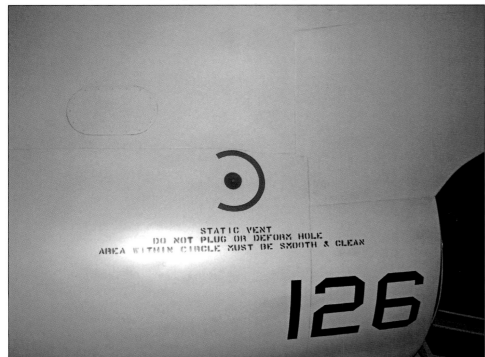

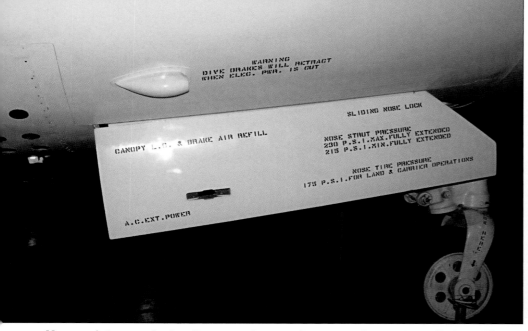

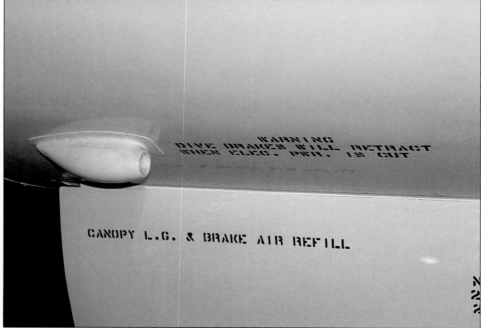

Many maintenance instructions, warnings, and access points are stenciled on different areas of the F9F Cougar. (Ken Neubeck)

Located just above the right side of the nose landing gear door is a warning to the ground crew regarding the dive brakes retraction when power is cut off. (Ken Neubeck)

Located below the left side of the canopy is the oxygen refill access, with quick-release latch and a warning to check cables when closing the door. (Ken Neubeck)

A similar oxygen refill access panel is located on the right side of the canopy as well, although without stenciled markings and caution warnings. (Ken Neubeck)

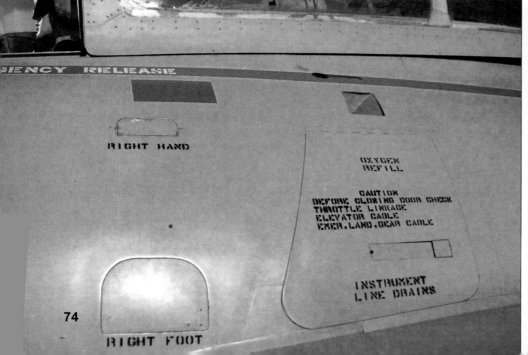

There are gravity fuel-cell fill points at different locations on the F9F Cougar. This fill point, surrounded by various warnings, is on the top of the left wing. (Ken Neubeck)

There are also gravity fill points for the fuselage fuel cells that are located on the top of the fuselage, behind the canopy. (Ken Neubeck)

Located on the lower fuselage, behind the nose landing gear, is an access port for pressure style refueling. (Ken Neubeck)

Instructions written on F9F fuselages vary from aircraft to aircraft. This F9F-8 carries a warning regarding the cannon. (Ken Neubeck)

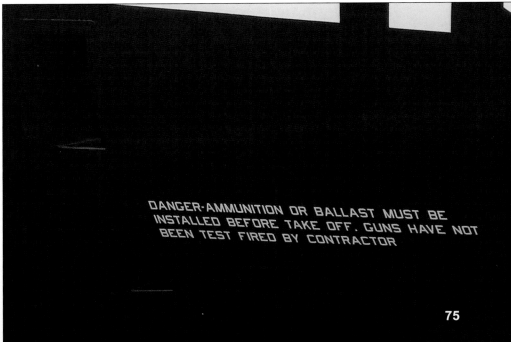

Fixed to the pylon is a bomb rack assembly that itself has two attachment points for ordnance, in this case a rocket launcher. (Ken Neubeck)

Underneath the lower fuselage of the Cougar, behind the air brakes, is a hook assembly that is used for catapult launching of the aircraft from carriers. (Ken Neubeck)

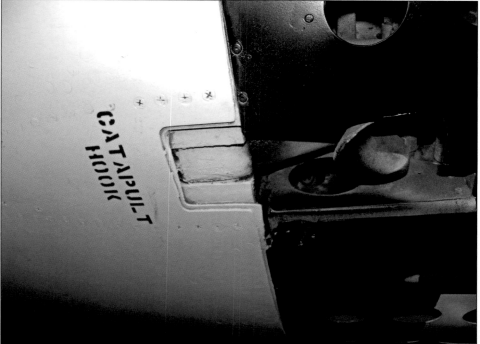

Early F9F aircraft, such as the F9F-6 and the F9F-7 models, were fitted with one underwing pylon on each wing that could be used to carry armaments. The configuration above shows a dual rocket launcher that is attached to the underwing pylon. This particular rocket, however, was never used on the F9F aircraft and is shown for demonstration purposes only. Later models such as the F9F-8 series would have additional underwing pylons added to both wings, in order to carry ordnance such as the AIM sidewinder missiles. (Ken Neubeck)

An F9F-8P Cougar receives fuel from AD-6 Skyraider during in-flight refueling over the ocean in 1958. (U.S. Navy)

This F9F-8P Cougar is about to be catapulted from the deck of the aircraft carrier *Franklin D. Roosevelt* (CVA-42) in 1958. (U.S. Navy)

For a few years in the 1950s, F9F-8 Cougars were flown by the Blue Angels. Two F9F-8T Cougars flew for a longer time with them. (Grumman archives via Cradle of Aviation)

With the Argentine Navy anchor motif on its wings, this F9F-8T, serial number 3-A-152, is one of two Cougars that were obtained by Buenos Aires. (Adolfo Soto)

77

This orange F9F, BuNo 128100, is one of the retired Cougars that were converted to QF-9G target drones in the 1960s. (Clay Jansson via Tailhook Association)

Another ex-F9F-6 model is this QF-9G, BuNo 128152 which has orange paint scheme with green tail color variation. (Clay Jansson via Tailhook Association)

This F9F-6KD, BuNo 130944, had multiple paint schemes with an orange-and-red fuselage and a yellow rudder. (Clay Jansson via Tailhook Association)

Like this former F9F-8 Cougar, BuNo 141194, later target drone aircraft were designated QF-9J. (Clay Jansson via Tailhook Association)

F9F-8, BuNo 131095, from attack squadron VA-66, sits on a carrier deck as the USS *Wisconsin* battleship steams off to the side in the background. (U.S. Navy)

QF-9G, BuNo 128100, has orange paint scheme with green tail color variation along with tiger's mouth motif on nose section. (Clay Jansson via Tailhook Association)

Its canopy slid open, F9F-9 BuNo 127413 sits parked on a carrier deck during the late 1950s. (Grumman archives via Cradle of Aviation)

On display at the Base Aeronaval Comandante Espora in Argentina, F9F-8T, S/N 3-A-151, was first Argentine aircraft to break the sound barrier. (Adolfo Soto)

An F-4B, BuNo 148403, on the left, and an F9F-8 Cougar, BuNo144283, on the right, sit on the field at NAS Oceana at Virginia Beach in 1964. Several new jets had been developed and were taking over the Cougar's fighter role by the 1960s, allowing the Navy to phase the Cougar out of active service. The insignia of Felix the Cat with a fuse bomb, along with top hat and skull and crossbones associated with this NAS are painted on the roof of the building. (Del Laughery)